THE POETRY OF LAWRENCIUM

The Poetry of Lawrencium

Walter the Educator

Silent King Books

Copyright © 2024 by Walter the Educator

All rights reserved. No part of this book may be reproduced in any manner whatsoever without written permission except in the case of brief quotations embodied in critical articles and reviews.

First Printing, 2024

Disclaimer
This book is a literary work; poems are not about specific persons, locations, situations, and/or circumstances unless mentioned in a historical context. This book is for entertainment and informational purposes only. The author and publisher offer this information without warranties expressed or implied. No matter the grounds, neither the author nor the publisher will be accountable for any losses, injuries, or other damages caused by the reader's use of this book. The use of this book acknowledges an understanding and acceptance of this disclaimer.

"Earning a degree in chemistry changed my life!"
— Walter the Educator

dedicated to all the chemistry lovers, like myself, across the world

SILENT KING BOOKS

SKB

Lawrencium's tale, a mystery untold,

LAWRENCIUM

In the depths of science, its wonders unfold.

LAWRENCIUM

Atomic number one zero three,

LAWRENCIUM

Rare and fleeting, a sight to see.

LAWRENCIUM

Born in fusion, stars collide,

LAWRENCIUM

In cosmic cauldrons, where elements bide.

LAWRENCIUM

Named for Lawrence, a Nobel mind,

LAWRENCIUM

In labs of wonder, its traces find.

LAWRENCIUM

Transuranic element, heavy and bright,

LAWRENCIUM

In the quest for knowledge, it takes flight.

LAWRENCIUM

A fleeting glimpse in the scientist's gaze,

LAWRENCIUM

Lawrencium's glow, a spectral blaze.

LAWRENCIUM

Unraveling atoms, probing the core,

LAWRENCIUM

In the quest for truth, we explore more.

LAWRENCIUM

In cyclotrons, with beams that race,

LAWRENCIUM

Lawrencium's atoms find their place.

LAWRENCIUM

A fleeting moment, a fraction of time,

LAWRENCIUM

In the dance of particles, so sublime.

LAWRENCIUM

Yet in its brevity, a story lies,

LAWRENCIUM

Of cosmic forces, where science flies.

LAWRENCIUM

Lawrencium's legacy, beyond compare,

LAWRENCIUM

In the realm of elements, it takes its share.

LAWRENCIUM

Its electrons dance in a quantum trance,

LAWRENCIUM

In shells of mystery, they seem to prance.

LAWRENCIUM

Unstable, fleeting, yet full of might,

LAWRENCIUM

Lawrencium's essence, a beacon of light.

LAWRENCIUM

In the depths of the periodic chart,

LAWRENCIUM

Lawrencium's presence, a work of art.

LAWRENCIUM

A testament to nature's boundless design,

LAWRENCIUM

In the realm of elements, it's a sign.

LAWRENCIUM

So let us ponder, in awe and wonder,

LAWRENCIUM

Lawrencium's journey, lightning and thunder.

LAWRENCIUM

A fleeting glimpse of the cosmic dance,

LAWRENCIUM

In the tapestry of creation, we take our chance.

LAWRENCIUM

In labs and chambers, where science reigns,

LAWRENCIUM

Lawrencium's secrets, the mind refrains.

LAWRENCIUM

Yet in its essence, a spark ignites,

LAWRENCIUM

In the human quest for truth's delights.

LAWRENCIUM

So raise a glass to Lawrencium's name,

LAWRENCIUM

In the annals of science, it claims its fame.

LAWRENCIUM

ABOUT THE CREATOR

Walter the Educator is one of the pseudonyms for Walter Anderson. Formally educated in Chemistry, Business, and Education, he is an educator, an author, a diverse entrepreneur, and he is the son of a disabled war veteran. "Walter the Educator" shares his time between educating and creating. He holds interests and owns several creative projects that entertain, enlighten, enhance, and educate, hoping to inspire and motivate you.

Follow, find new works, and stay up to date
with Walter the Educator™
at WaltertheEducator.com

www.ingramcontent.com/pod-product-compliance
Lightning Source LLC
LaVergne TN
LVHW051921060526
838201LV00060B/4111